5 Steps Chemistry Workbook Series
Book 2: Mighty Mole Concepts

14 Worksheets and Keys for College Prep and Honors Chemistry High School Students

By Julie C. Gilbert

Aletheia Pyralis Publishers

For information about special discounts available for bulk purchases, sales promotions, fund-raising and educational needs, please email: juliecgilbert5steps@gmail.com or devyaschildren@gmail.com (this one is checked more often).

http://www.juliecgilbert.com/

Table of Contents:

Introduction:

Note: If you've read *5 Steps Chemistry Workbook Series Book 1: Sassy Stoichiometry Problems*, you will note that most of the introduction is the same. That's because I intend to structure the book exactly the same way. However, the "Notes about the Worksheets" section may differ. This book will have more sections than stoichiometry because the topic is broken down into many more subtopics than stoichiometry.

Second Note: I'm aware Mole Concepts come first in the curriculum, but the series order was established by when they were written.

What is this book and how do I use it?

This will be a collection of worksheets designed for college prep and honors chemistry students to practice mole concepts. Each section will have the "easier" practice sets first.

For convenience, I will likely be providing a list of the atomic masses that will be used in this book because every periodic table will round to a different number of decimal places. For testing and quizzing purposes, you will likely have to use the periodic table provided by your teacher.

Tip: If your teacher allows it, you may wish to write out a list of common atomic masses you will need. With enough practice, you'll likely memorize the important ones, but there are several elements that show up in a lot of compounds: hydrogen, oxygen, carbon, nitrogen, sulfur, phosphorus.

Tip: My keys and explanations will be designed around one of the most basic periodic tables I can find because I'd rather not drown in decimals along the way. However, you should solve the problems with your class periodic table. I also recommend you use the calculator you intend to use during assessments. The more you use the normal tools at your disposal, the more comfortable you will be with the process.

What are the Mole Concepts?

Your teacher might have broken the topic down differently.

I break Mole Concepts down this way:

- Formula mass and molar mass calculations
- Mole conversions – this includes mole to mass, mole to particle (and vice versa) calculations, and two-step conversions
    - **Side note:** mole of a compound to mole(s) of a particular element within that compound may be discussed, but most people save mole to mole conversions for stoichiometry.
- Percent calculations – percent composition of a compound, percent of water in a hydrate
- Empirical and Molecular Formulas
- Calculation of the number of water molecules in a hydrate

Note: Percent of water in a hydrate assumes you already know what the hydrate is and want to figure out how much of the molecule's mass consists of water. Calculation of the number of water molecules in a hydrate uses the rest of mole concepts to help you determine the chemical formula for a hydrate. If that's about as clear as mud now, don't worry. The individual worksheets should clarify the differences in the two tasks.

Why should I use this book when I can find a million articles and a dozen videos online?

When it comes to mastering chemistry concepts, there's really no substitute for practice. Please do find some videos that cover the topic and use them to guide you. This will be a book of worksheets with the keys available throughout. The keys will be more than just answers. They should provide both the mathematical work and written explanations for how I got there. Most worksheets will have been student tested.

Translation: I'm going to subject my current class to a wide variety of questions from here, so we can catch all those pesky calculation errors that crop up when these things happen in isolation.

I absolutely love audiobooks and ebooks are okay too, but for this, I highly recommend the paperback because the main point is for each student to be able to physically write out the answers.

Notes about the Worksheets in this Book:

Note 1: Most of the time I will use the chemical formulas instead of the name, but occasionally, I will switch it up. If you struggle with writing chemical formulas, you should probably practice that concept until mastery. Who knows, maybe someday, I'll have another workbook for you, but that day is not today. I may also write both during the training worksheets, but your teacher may use only names or only formulas, depending on the class level.

Note 2: You will sometimes need to use exponents. I do not let my students write E in place of x10^, but your teacher might. However, when it comes to plugging numbers in a calculator, you will likely want to use the EE button. This is often a second function on your calculator. Mine (a TI-30XIIS) has it above X^{-1}.

Note 3: The answer keys will be immediately after each worksheet.

Note 4: The typed fractions do not format the subscripts properly. You should use proper subscripts when writing chemical formulas.

Note 5: I will sometimes change the wording of the questions and explanations because there's no way to predict the way your teacher will write questions, and sometimes, a particular wording will work for a student where another one fails to connect. It's best if you see a variety of ways to word problems and explanations. For this reason, some explanations may be presented as bullet points.

Note 6: For the same reason as note 5, I will occasionally switch up the number of significant figures. Always read the directions in a particular worksheet. If your teacher doesn't care about significant figures, then count your lucky stars and keep a reasonable (means 2-3) number of decimals in your answer.

Crash Course in significant figures rules:
Non-zero digits = significant
Zeroes sandwiched between non-zero digits = significant
Zeroes following a decimal point and a real digit = significant. (For example, .0002 has one significant figure, but 1.0002 has five significant figures.)
Trailing zeroes are not significant unless there is a decimal point present after. (For example, 2000 has one significant figure, but 2000. has four significant figures.)

Note 7: For empirical and molecular formulas, write the chemicals in the order that the percents or grams is given. (Yes, percents probably isn't a word. Probably should be percentages, but I kind of like percents.)

Note 8: General Directions: Keep all units and labels throughout each problem. Watch significant figures unless otherwise instructed.

Note 9: Here are the most common element masses used in the book. I occasionally use other elements. If I have not mentioned it here, you can look up the molar mass and round to the nearest 1 decimal.

H = 1.0	Na = 23.0	S = 32.1	Cr = 52.0
C = 12.0	Mg = 24.3	Cl = 35.5	Fe = 55.8
N = 14.0	Al = 27.0	Ca = 40.1	Au = 197.0
O = 16.0	P = 31.0	Ti = 47.9	Pb = 207.2

Overview, Background, Common Mistakes:

Disclaimer: Most of the basic explanatory text is the same found in *5 Steps to Surviving Chemistry*. I see no reason to reinvent the wheel I've already made. However, the structure is going to be different because I'm going to break it up with the relevant worksheets.

Quick Review of Conversions:

Step #1: Start with given information
Step #2: Put units for the unit that needs to change on the bottom and units that you're changing to on the top
Step #3: Repeat step 2 until you get to the final unit
Step #4: Plug in the conversion factors
Step #5: Solve (top = multiply, bottom = divide)

Most Common Mistakes:

- **Not using units!** Units are your friends. Allow them to help you convert from one substance to another.
- Units that don't match. If you want to cancel a unit, you must have the same unit (or unit/label combination)
  How many hours are in two years?
  $2 \text{ years x } \frac{365\ days}{1\ year} \text{ x } \frac{24\ hours}{1\ day} = 17520 \text{ hours}$
- Improper abbreviations. Formula units = form u (no exceptions; fu is just not polite); moles = mol; molecules = molecules; atoms = atoms; grams = g; grams per mole = g/mol (not gm or gpm or whatever else your imaginations can come up with)

Mole Concepts Introduction:

This chapter contains a lot of different topics, but they all revolve around the mole. It's probably going to be a long chapter so skim the titles of each section if you're just looking for something specific.

Definitions:

Formula unit vs. molecule:
Formula unit: the simplest ratio of ions in an ionic compound
Molecule: made of two or more atoms that are covalently bound to one another
Note: same idea, different term based on the type of compound you're dealing with

Use formula unit when dealing with ionic compounds.
Use molecule when dealing with covalent compounds.

Atomic mass units (amu) vs. grams (g):
For all intents and purposes, they are equivalent units of mass.
Note: you might see it referred to as u for unified atomic mass unit or Da for Dalton. There's a long history of naming units after famous dead guys (notable scientists).

Atomic mass units: one atomic mass unit is 1/12 of the mass of a Carbon-12 atom
Note: There is a technical difference in the mass of amu and grams, but in the amount of significant figures we deal with on the periodic table the difference is negligible (too small to impact the results).

Formula mass vs. Molar mass:
Formula mass: the mass in one formula unit of a substance
Molar mass: the mass per mole of a substance

The average atomic mass found on the periodic table is usually accepted as the mass per mole of that element.

Steps to Calculating Formula Mass and Molar Mass:

Step #1: Write the balanced chemical formula for the compound
Step #2: Write down the number of each type of atom
Step #3: Convert from the number of atoms to the mass by using the average atomic mass found on the periodic table of elements. (The number of atoms goes on the bottom of the conversion factor; mass of the element goes on top of the conversion factor.)
Step #4: Add the values together and report the final answer.
Note: There's a long and a short form of this concept. I'll show you the long form once so you can see what's going on, but then I usually don't ask to see long form.
Note: Molar mass is obtained the same way.

Example: What is the formula mass for calcium chloride? (Long form)
Step #1: $CaCl_2$
Step #2: 1 atom of Ca and 2 atoms of Cl
Step #3:
1 Ca atom x $\frac{40.1\ amu}{1\ Ca\ atom}$ = 40.1 amu
2 Cl atoms x $\frac{35.5\ amu}{1\ atom\ Cl}$ = 71.0 amu
Step #4: 40.1 amu + 71.0 amu = 111.1 amu

Example: What is the formula mass for calcium chloride? (Short form)
Step #1: $CaCl_2$
Step #2: 1 atom of Ca and 2 atoms of Cl
Step #3:
1 Ca x 40.1 amu = 40.1 amu
2 Cl x 35.5 amu = 71.0 amu
Step #4: 40.1 amu + 71.0 amu = 111.1 amu

Name: ______________________________ Date: ______________ Period: ______

Worksheet 1: Formula Mass and Molar Mass Calculations

Directions: Find the formula mass or molar mass as indicated. For simplicity, you can ignore sig figs here. Just keep all decimals.

Note: Some of you probably need the practice putting your name on stuff.
Second Note: Short form is okay.

1. What is the formula mass of MgI_2 (magnesium iodide)?

2. What is the formula mass for Co_2O_3 (cobalt (III) oxide)?

3. What is the formula mass for $BaCl_2$ (barium chloride)?

4. What is the molar mass of $(NH_4)_3P$ (ammonium phosphide)?

5. What is the molar mass of $Sc_2(SO_4)_3$ (scandium sulfate)?

Name: ______________________________ Date: ____________ Period: _____

KEY for Worksheet 1: Formula Mass and Molar Mass Calculations

Directions: Find the formula mass or molar mass as indicated. For simplicity, you can ignore sig figs here. Just keep all decimals.

Note: Some of you probably need the practice putting your name on stuff.
Second Note: Short form is okay.

1. What is the formula mass of MgI_2 (magnesium iodide)?

1 Mg x 24.3 amu = 24.3 amu
2 I x 126.9 amu = <u>253.8 amu</u>
278.1 amu

Explanation: The formula for magnesium iodide contains 1 Mg and 2 I. The mass of Mg is 24.3 amu. When multiplied by the 1, it is still 24.3 amu. The mass of I is 126.9 amu. When multiplied by 2 because there are two I's in the formula, it becomes 253.8 amu. The final answer is obtained by adding both of these values: 24.3 amu + 253.8 amu = 278.1 amu.

2. What is the formula mass for Co_2O_3 (cobalt (III) oxide)?

2 Co x 58.9 amu = 117.8 amu
3 O x 16.0 amu = <u>48.0 amu</u>
165.8 amu

Explanation: The formula for cobalt (III) oxide contains 2 Co and 3 O. The mass of Co is 58.9 amu. When multiplied by 2, this becomes 117.8 amu. The mass of O is 16.0 amu. When multiplied by 3 because there are three of the O atoms, this becomes 48.0 amu. The formula mass is found by adding these two values: 117.8 amu + 48.0 amu = 165.8 amu.

3. What is the formula mass for $BaCl_2$ (barium chloride)?

1 Ba x 137.3 amu = 137.3 amu
2 Cl x 35.5 amu = <u>71.0 amu</u>
208.3 amu

Explanation: The formula for barium chloride contains 1 Ba and 2 Cl. Barium (Ba) has a mass of 137.3 amu. The contribution from Ba to the formula mass will be 137.3 amu because there is only 1 in the formula. There are 2 Cl atoms in the formula. Therefore, the mass contribution from Cl will be 2 x 35.5 amu = 71.0 amu. The total formula mass will be the sum of the contributions from Ba and Cl: 137.3 amu + 71.0 amu = 208.3 amu.

4. What is the molar mass of $(NH_4)_3P$ (ammonium phosphide)?

3 N x 14.0 g = 42.0 g
12 H x 1.0 g = 12.0 g
1 P x 31.0 g = 31.0 g
85.0 g/mol $(NH_4)_3P$

Explanation: 1 mole of ammonium phosphide contains 3 moles of N, 12 moles of H, and 1 mole of P. To get that far, you will have to remember to distribute the subscript 3 to everything inside the parentheses. 1 mole of N has a mass of 14.0 g. Because there are 3 N's in the formula, the contribution from N overall is 42.0 g. Each mole of H contributes 1.0 g to the mass. Since there are 12 H's, the mass contribution is 12.0 g. There is only 1 P in the formula and it has a mass of 31.0 g. These subanswers are added to get the molar mass of $(NH_4)_3P$. 42.0 g + 12.0 g + 31.0 g = 85.0 g/mol $(NH_4)_3P$.

Please note: We do not see the moles disappear in the conversions because we have opted to do short form. The final unit still contains mol because it is referring to the entire substance: 1 mole of $(NH_4)_3P$.

5. What is the molar mass of $Sc_2(SO_4)_3$ (scandium sulfate)?

2 Sc x 45.0 g = 90.0 g
3 S x 32.1 g = 96.3 g
12 O x 16.0 g = 192.0 g
378.3 g/mol $Sc_2(SO_4)_3$

Explanation: This problem is solved the same way as the others. A mole of $Sc_2(SO_4)_3$ contains 2 moles of Sc, each of which has a mass of 45.0 g, so the mass contribution is 90.0 g. It also contains 3 moles of S. The mass of each S is 32.1 g, so 3 x this is 96.3 g. There are 12 moles of O, each of which has a mass of 16.0 g, so the mass contribution from O is 192.0 g. Adding the subanswers will give the total molar mass: 90.0 g + 96.3 g + 192.0 g = 378.3 g/mol $Sc_2(SO_4)_3$.

Mole Conversions:

What is a mole?

The mole is a very important unit. It's been called the chemist's dozen. At this point in your academic career, you should be familiar with the concept of a dozen. A dozen is twelve items. A mole is 6.022×10^{23} items. In chemistry, those items are called atoms, formula units, molecules, ions, or simply particles if you want to generalize more.

Moles allow for conversions between the number of particles and the mass of those particles. Some students have a hard time grasping numbers as large as 6.022×10^{23}. So, let's return to the concept of a dozen to grasp the relationship to mass. If I have a dozen (twelve) identical pencils, they will have a certain mass. You could measure the mass on a balance. If I have a dozen (twelve) identical cars, they too will have a certain mass. If we had a large enough scale, we could measure the mass of the cars.

The number of the items in both cases is a dozen (twelve), but the masses will be vastly different. The same holds true for atoms, ions, and other particles. A hydrogen atom has a mass of 1.00 g. A gold atom has a mass of 197.0 g. So, a mole of hydrogen atoms and a mole of gold atoms would both contain 6.022×10^{23} number of atoms, but the masses of those would be vastly different.

Steps to mole conversions and using molar mass as a conversion factor:

Step #1: Write down the number, unit, and label for the starting compound.
Step #2: Put the unit that matches the starting compound on the bottom of the conversion.
Step #3: Put the unit that you'd like to change to on the top of the conversion.
Step #4: Repeat steps 3 and 4 until you've changed completely over to the desired unit and label.
Step #5: Solve and check your answer.

Note 1: It's the same steps for formula mass being used as conversions, just with different units.
Note 2: By "label" I mean the chemical formula.
Note 3: For obvious reasons, I don't let my students completely abbreviate formula units. (It's just not polite.) I make them write form u if they want to abbreviate it.

Example 1: a one-step problem going from formula units to moles
How many moles are equal to 2.55×10^{28} formula units of lithium carbonate?
Step #1: 2.55×10^{28} formula units Li_2CO_3
Step #2: Put the unit that matches the starting compound on the bottom of the conversion.

$$2.55 \text{ x}10^{28} \text{ form units } Li_2CO_3 \text{ x } \frac{\quad}{6.02 \, x \, 10^{23} form \, u \, Li2CO3}$$

Step #3: Put the unit that you'd like to change to on the top of the conversion.

$$2.55 \text{ x}10^{28} \text{ form units } Li_2CO_3 \text{ x } \frac{1 \, mol \, Li2CO3}{6.02 \, x \, 10^{23} form \, u \, Li2CO3} =$$

Step #4: Repeat steps 3 and 4 until you've changed completely over to the desired unit and label.
Not necessary here since I'm already in the unit I want to be in after this first conversion.

Step #5: Solve and check your answer.

2.55×10^{28} form units Li_2CO_3 x $\frac{1\ mol\ Li2CO3}{6.02\ x\ 10^{23} form\ u\ Li2CO3}$ = 42358.8039867 mol Li_2CO_3

To three sig figs, the answer would be 42400 mol Li_2CO_3 or 4.24×10^4 mol Li_2CO_3.

Example 2: a two-step problem going from formula units to moles and then to grams
How many g are equal to 2.55×10^{28} formula units of lithium carbonate?
Step #1: 2.55×10^{28} formula units Li_2CO_3
Step #2: Calculate the molar mass of the compound and set up a conversion.
The molar mass of Li_2CO_3 = 73.8 g/mol
Step #3: Put the unit that matches the starting compound on the bottom of the conversion.

2.55×10^{28} form units Li_2CO_3 x $\frac{\quad}{6.02\ x\ 10^{23} form\ u\ Li2CO3}$

Step #4: Put the unit that you'd like to change to on the top of the conversion.

2.55×10^{28} form units Li_2CO_3 x $\frac{1\ mol\ Li2CO3}{6.02\ x\ 10^{23} form\ u\ Li2CO3}$

Step #5: Repeat these steps until you've changed completely over to the desired unit and label.

2.55×10^{28} form units Li_2CO_3 x $\frac{1\ mol\ Li2CO3}{6.02\ x\ 10^{23} form\ u\ Li2CO3}$ x $\frac{73.8\ g\ Li2CO3}{1\ mol\ Li2CO3}$

Step #6: Check your answer.

2.55×10^{28} form units Li_2CO_3 x $\frac{1\ mol\ Li2CO3}{6.02\ x\ 10^{23} form\ u\ Li2CO3}$ x $\frac{73.8\ g\ Li2CO3}{1\ mol\ Li2CO3}$ = 3.13×10^6 g Li_2CO_3

Reminder: Compounds are created from elements combined in ratios of small whole numbers.
Tip: The following is something I have all my students write on a notecard. If it helps, great. If it confuses you, ignore it. This is a summary of the types of conversions you will need to do.

$\frac{1\ mole}{g}$ $\frac{6.022\ x\ 10^{23} particles}{1\ mole}$

---------------> ------------------>

Mass ← molar mass → Moles ← Avogadro's Number → Particles

<--------------- <-----------------

$\frac{g}{1\ mole}$ $\frac{1\ mole}{6.022\ x\ 10^{23} particles}$

Explanation:
You could be asked to convert from mass to moles, moles to particles, or mass to particles, or be asked to go the other way from particles to moles, moles to mass, or particles to mass. The card should help you figure out which conversion to use and what it looks like.

Going from mole of a compound to moles of just a part of the compound:

These problems would exist in the middle of the above chart.

I do this type of problems some years and skip it others. Your teacher may not cover this topic because it's similar to Stoichiometry, which is the next logical unit in most chemistry curricula.

It springs from the fact that a mole is like a dozen. It just means a certain number of some item or in our case, any type or portion of a chemical. Recall that particles can refer to atoms, ions, formula units, or molecules.

For example, 1 mole of $CaSO_4$ contains 1 mole of Ca, 1 mole of S, and 4 moles of O. Even if you are asked to go to something where the mole ratio between compound and element is 1:1, you will need to do this step because it's often accompanied by asking you to go to grams of something.

Sample problem:
What mass in grams of oxygen is contained in 9.57×10^{24} formula units of $CaSO_4$?

$$9.57 \text{ x } 10^{24} \text{ form u } CaSO_4 \text{ x } \frac{1\ mol\ CaSO4}{6.02\ x\ 10^{23} form\ u\ CaSO4} \text{ x } \frac{4\ mol\ O}{1\ mol\ CaSO4} \text{ x } \frac{16.0\ g\ O}{1\ mol\ O} = 1020 \text{ g O}$$

Explanation: Start with the given, which is 9.57×10^{24} formula units of $CaSO_4$. Set up three conversion factors. The first will have Avogadro's number of formula units under 1 mol of $CaSO_4$. This gets us to moles of the entire chemical. Next, we get out of moles of the entire chemical and into moles of just the part we're interested in, the oxygen. This conversion has 4 mols O on top and 1 mol of $CaSO_4$ on the bottom. The last conversion gets us out of moles of oxygen and into grams of O.

Keeping the chemical labels is annoying but necessary because the middle conversion has the same unit. The last molar mass depends on the identity of the atom we went to, in this case O.

The exact answer spit out by my calculator was 1017.408638 g O. I rounded to 3 significant figures because that's very common for chemistry teachers to ask you to do.

Note: There is an alternate way to solve this problem if you are very comfortable using percent calculations. Percents will be explained in more detail later. You may wish to return to this section after studying percent calculations.

Second note: This way is slightly longer in this case because we are not given the grams of $CaSO_4$ to begin with.

Alternate Solution using percents:

$$9.57 \text{ x } 10^{24} \text{ form u } CaSO_4 \text{ x } \frac{1\ mol\ CaSO4}{6.02\ x\ 10^{23} form\ u\ CaSO4} \text{ x } \frac{136.2 \text{ g } CaSO4}{1\ mol\ CaSO4} = 2165 \text{ g } CaSO_4$$

You need to get to the grams of the entire compound. This isn't our final answer, so it's good not to round in this situation.

% of O in the compound:
1 Ca x 40.1 g = 40.1 g
1 S x 32.1 g = 32.1 g

$4\ O \times 16.0\ g = +\underline{64.0\ g}$

$136.2\ g/mol\ CaSO_4$

$$\%\ O: \frac{64.0\ g\ O}{136.2\frac{g}{mol}\ CaSO4} \times 100 = 46.99\%$$

You do not need the entire percent composition, just the percent of oxygen. You don't even have to take it to a full percent. In fact, you want the decimal percent.

Multiply the grams of the entire compound by the decimal percent of O to get your final answer.

$2165\ g\ CaSO_4 \times .4699 = 1020\ g\ O$

The answer should be identical to what you got before. This time I got 1017.3335 g O.

The answers aren't exact, but once rounded to significant figures, they should be the same. The difference comes in because of rounding we did along the way.

Name: ________________________________ Date: _______________ Period: ______

Worksheet 2: Moles to Particles and Particles to Moles Conversions

Directions: Use the conversion factor method to solve each of the problems.

Note: Particles is a catch-all term that can include atoms, molecules, formula units, and even ions depending on the context.

Second Note: Avogadro's number is 6.02×10^{23}, so this is the equivalent number of particles per mole of substance. There are other versions of it that preserve more decimals.

Third Note: I may use significant figures, sig figs, and SF interchangeably.

1. How many moles of magnesium are equal to 3.45×10^{24} atoms of Mg?

2. How many moles of sodium chloride are equal to 4.54×10^{22} formula units of NaCl?

3. How many formula units of potassium hydroxide are equal to 3.50 moles of KOH?

4. How many molecules of water are equal to .684 moles of H_2O?

5. How many atoms of vanadium are contained in 6.51 moles of V?

Name: ________________________________ Date: _______________ Period: ______

KEY for Worksheet 2: Moles to Particles and Particles to Moles Conversions

Directions: Use the conversion factor method to solve each of the problems.

Note: Particles is a catch-all term that can include atoms, molecules, formula units, and even ions depending on the context.

Second Note: Avogadro's number is 6.02×10^{23}, so this is the equivalent number of particles per mole of substance. There are other versions of it that preserve more decimals.

Third Note: I may use significant figures, sig figs, and SF interchangeably.

1. How many moles of magnesium are equal to 3.45×10^{24} atoms of Mg?

$$3.45 \times 10^{24} \text{ atoms Mg} \times \frac{1\ mol\ Mg}{6.02\ x\ 10^{23} atoms\ Mg} = 5.73 \text{ moles Mg}$$

Explanation: Start with the given, which is 3.45×10^{24} atoms Mg. Set up a conversion factor to get out of atoms and into moles. This would require Avogadro's number (6.02 x 10^{23} of Mg atoms) to be on the bottom part of the fraction and 1 mol Mg to be on the top part. Solve by dividing 3.45×10^{24} atoms Mg by 6.02×10^{23} of Mg atoms to get 5.73 moles of Mg. The significant figures for the answer are 3 because of the given information having 3 digits.

2. How many moles of sodium chloride are equal to 4.54×10^{22} formula units of NaCl?

$$4.54 \times 10^{22} \text{ formula units NaCl} \times \frac{1\ mol\ NaCl}{6.02\ x\ 10^{23} formula\ units\ NaCl} = .0754 \text{ mol NaCl}$$

Explanation: Start with the given, which is 4.54×10^{22} formula units NaCl. The conversion for this looks identical to the previous problem because we are still given a particle type and asked to change into moles. The difference in units is due to the difference between Mg (an atom) and NaCl (an ionic compound). Divide 4.54×10^{22} formula units NaCl by 6.02×10^{23} formula units of NaCl. This results in .0754 mol NaCl. Some people put a 0 in front of the decimal. That is fine.

3. How many formula units of potassium hydroxide are equal to 3.50 moles of KOH?

$$3.50 \text{ moles KOH} \times \frac{6.02\ x\ 10^{23} formula\ units\ KOH}{1\ mol\ KOH} = 2.11 \times 10^{24} \text{ formula units KOH}$$

Explanation: Start with the given, which is 3.50 moles KOH. To convert to formula units, that has to be on the top part of the fraction. 1 mole KOH is what you are canceling, so that goes on the bottom. When the multiplication is done and you round to 3 SF, you get 2.11×10^{24} formula units KOH.

4. How many molecules of water are equal to .684 moles of H_2O?

.684 moles H_2O x $\frac{6.02 \times 10^{23} molecules\ H2O}{1\ mol\ H2O}$ = 4.12×10^{23} molecules H_2O

Explanation:

- Start with the given, which is .684 moles of H_2O.
- The conversion factor will have 6.02×10^{23} molecules H_2O on top and 1 mol H_2O on the bottom.
- Solve mathematically by multiplying the two values.
- The answer is 4.12×10^{23} molecules H_2O.

5. How many atoms of vanadium are contained in 6.51 moles of V?

6.51 moles of V x $\frac{6.02 \times 10^{23} atoms\ V}{1\ mol\ V}$ = 3.92×10^{24} atoms V

Explanation:

- Start with the given, which is 6.51 moles of V.
- The conversion factor will have 6.02×10^{23} atoms V on top and 1 mol V on the bottom. This allows you to cancel the moles and end up in atoms.
- Mathematically, you will be multiplying to solve the problem.
- The answer is 3.92×10^{24} atoms V. We are using atoms because V is an element.

Name: ______________________________ Date: ______________ Period: ______

Worksheet 3: Moles to Mass and Mass to Moles Conversions

Directions: Use the conversion factor method to solve each of the problems.

Note: You will need to calculate the molar mass first. I am not going to cover that part in the explanations.

1. How many moles of calcium chloride are equal to 52.4 g of $CaCl_2$?

2. How many moles of vanadium (V) fluoride are equal to 6.89 g of VF_5?

3. How many grams of manganese (II) chromate are equal to .935 moles of $MnCrO_4$?

4. How many grams of carbon dioxide are equal to 7.8 moles of CO_2?

5. How many grams of rubidium carbonate are equal to .500 moles of Rb_2CO_3?

Name: ______________________________ Date: ______________ Period: ______

Key for Worksheet 3: Moles to Mass and Mass to Moles Conversions

Directions: Use the conversion factor method to solve each of the problems.

Note: You will need to calculate the molar mass first. I am not going to cover that part in the explanations.

1. How many moles of calcium chloride are equal to 52.4 g of $CaCl_2$?

$$52.4 \text{ g } CaCl_2 \times \frac{1\ mol\ CaCl2}{111.1\ g\ CaCl2} = .472 \text{ mol } CaCl_2$$

Explanation: Start with the given, which is 52.4 g $CaCl_2$. Set up a conversion factor to get out of grams and into moles. It will have moles on top and grams on the bottom. The molar mass of $CaCl_2$ is 111.1 g/mol. Since grams is on the bottom, we divide. 52.4 g/ by 111.1 g = .472 mole $CaCl_2$ when rounded to 3 SF.

2. How many moles of vanadium (V) fluoride are equal to 6.89 g of VF_5?

$$6.89 \text{ g } VF_5 \times \frac{1\ mol\ VF5}{145.9\ g\ VF5} = .0472 \text{ mol } VF_5$$

Explanation: Start with the given, which is 6.89 g VF_5. Like the previous problem, you need to be in moles at the end, so that is the top part of the molar mass conversion. To get rid of the unit g, you will need to divide by the number of grams in a mole of VF_5. The answer is .0472 mol VF_5.

3. How many grams of manganese (II) chromate are equal to .935 moles of $MnCrO_4$?

$$.935 \text{ moles } MnCrO_4 \times \frac{170.9\ g\ \text{MnCr}O4}{1\ mol\ \text{MnCr}O4} = 160. \text{ g } MnCrO_4$$

Explanation: Start with the given, which is .935 moles $MnCrO_4$. This time you want to end up in grams, so that becomes the top part of the conversion fraction. The molar mass of $MnCrO_4$ is 170.9 g/mol. Because the 170.9 g is on the top part, we solve by multiplying this by the original starting value of .935 moles $MnCrO_4$. The moles portion goes away, leaving us with g as a unit. The numerical answer is 160. g $MnCrO_4$.

4. How many grams of carbon dioxide are equal to 7.8 moles of CO_2?

7.8 moles CO_2 x $\frac{44.0\ g\ CO2}{1\ mol\ CO2}$ = 340 g CO_2

Explanation:

- Start with the given, which is 7.8 moles CO_2.
- To convert out of moles and into grams, you will set up a conversion factor using the molar mass of CO_2, which is 44.0 g/mol. To cancel moles and end up in grams, the 44.0 g needs to be on top.
- When solving, you will multiply 7.8 moles CO_2 by 44.0 g CO_2.
- The bottom part of the fraction will have canceled the unit of moles.
- The answer is 340 g CO_2. Please note that this is only 2 SF.

5. How many grams of rubidium carbonate are equal to .500 moles of Rb_2CO_3?

.500 moles Rb_2CO_3 x $\frac{231\ g\ \mathrm{Rb2CO3}}{1\ mol\ \mathrm{Rb2CO3}}$ = 116 g Rb_2CO_3

Explanation:

- Start with the given, which is .500 moles Rb_2CO_3.
- Set up a conversion factor with g of 231 g Rb_2CO_3 on top and 1 mol Rb_2CO_3 on the bottom.
- To solve, multiply because the number is on top.
- The answer you get will be 116 g Rb_2CO_3 once rounded to significant figures.

Name: ______________________________ Date: ______________ Period: _____

Worksheet 4: 2-Step Mole Conversions

Directions: Use the conversion factor method to solve each of the problems.

Note: You will need to calculate the molar mass first. I am not going to cover that part in the explanations. These are solved like 1-step conversion problems, but you need to first convert to moles and then to the final unit.

Avogadro's number: 6.02×10^{23} particles/mol.

1. What is the mass (in g) of 6.44×10^{21} atoms of lithium (Li)?

2. How many moles of H are contained within 64.5 g C_6H_{12}?

3. How many molecules of glucose are needed to make 548 g of $C_6H_{12}O_6$?

4. How many formula units of aluminum hydroxide are contained in 64.5 g $Al(OH)_3$?

5. How many grams of carbon would be equal to 9.47×10^{25} atoms of C?

Name: ______________________________ Date: ______________ Period: _____

Key for Worksheet 4: 2-Step Mole Conversions

Directions: Use the conversion factor method to solve each of the problems.

Note: You will need to calculate the molar mass first. I am not going to cover that part in the explanations. These are solved like 1-step conversion problems, but you need to first convert to moles and then to the final unit.

Avogadro's number: 6.02×10^{23} particles/mol.

1. What is the mass (in g) of 6.44×10^{21} atoms of lithium (Li)?

$$6.44 \times 10^{21} \text{ atoms Li} \times \frac{1\ mol\ Li}{6.02 \times 10^{23} atoms\ Li} \times \frac{6.9\ g\ Li}{1\ mol\ Li} = .0738 \text{ g Li}$$

Explanation: Start with the given, which is 6.44×10^{21} atoms Li. Set up a conversion factor to get out of atoms and into moles. This requires Avogadro's number (6.02×10^{23} atoms Li) to be on the bottom and 1 mol Li to be on top. Next, convert to grams by putting 6.9 grams of Li on top and 1 mol Li on the bottom. Although you can do the math operations in any order, I tend to start at the beginning and do them step by step so I don't miss anything. I'd type in 6.44E21, hit divide, 6.02E23, hit enter, hit multiply, type 6.9 and hit enter again. The answer comes to .0738 g Li.

Note: In a calculator, E stands in place of x10^. This is how you get an exponent. On most calculators it's a second function and looks like EE.

2. How many moles of H are contained within 64.5 g C_6H_{12}?

$$64.5 \text{ g } C_6H_{12} \times \frac{1\ mol\ C6H12}{84\ g\ C6H12} \times \frac{12\ mol\ H}{1\ mol\ C6H12} = 9.21 \text{ g H}$$

Explanation: Since this is the portion of the workbook devoted to conversions, I will use that method for solving this problem. Start with the given, which is 64.5 g C_6H_{12}. Set up two conversions. The first conversion will have 1 mol C_6H_{12} on top and the grams of C_6H_{12} on the bottom (84 g C_6H_{12}). This is a molar mass conversion. The next conversion will take us out of 1 mol C_6H_{12} to just the portion we're interested in, the H. There are 12 moles of H in this compound. Multiply across the top and divide numbers on the bottom. The final answer is 9.21 g H

3. How many molecules of glucose are needed to make 548 g of $C_6H_{12}O_6$?

$$548 \text{ g } C_6H_{12}O_6 \times \frac{1\ mol\ C6H12O6}{180\ g\ C6H12O6} \times \frac{6.02 \times 10^{23} molecules\ C6H12O6}{1\ mol\ C6H12O6} = 1.83 \times 10^{24} \text{ molecules } C_6H_{12}O_6$$

Explanation: Start with the given, which is 548 g $C_6H_{12}O_6$. Use the molar mass of glucose as a conversion factor to get out of grams and into moles. This requires 1 mol $C_6H_{12}O_6$ to be on top and the 180 grams of glucose to be on the bottom. Next, get to molecules by using Avogadro's number. The 6.02×10^{23} molecules goes on top, and the 1 mol $C_6H_{12}O_6$ goes on the bottom. Solve by multiplying by numbers on the top parts of the fractions and dividing by numbers on the bottom parts of the fractions.

4. How many formula units of aluminum hydroxide are contained in 64.5 g $Al(OH)_3$?

$$64.5\text{ g Al(OH)}_3 \times \frac{1\,mol\text{ Al(OH)3}}{78\,g\,Al(OH)3} \times \frac{6.02 \times 10^{23} form\,u\text{ Al(OH)3}}{1\,mol\text{ Al(OH)3}} = 4.98 \times 10^{23}\text{ form u Al(OH)}_3$$

Explanation:

- Start with the given, which is 64.5 g $Al(OH)_3$.
- The first task is to get out of grams of $Al(OH)_3$ and into moles. 1 mol $Al(OH)_3$ will be on the top and 78 g $Al(OH)_3$ will be on the bottom of the first conversion factor.
- Next, get to formula units by putting 6.02×10^{23} form u $Al(OH)_3$ on top and 1 mol $Al(OH)_3$ on the bottom of the second conversion.
- Solve by starting with 64.5 dividing by 78 and multiplying by 6.02×10^{23}.
- The answer comes to 4.98×10^{23} form u $Al(OH)_3$.

5. How many grams of carbon would be equal to 9.47×10^{25} atoms of C?

$$9.47 \times 10^{25}\text{ atoms C} \times \frac{1\,mol\,C}{6.02 \times 10^{23} atoms\,C} \times \frac{12.0\,g\,C}{1\,mol\,C} = 1890\text{ g C}$$

Explanation:

- Start with the given, which is 9.47×10^{25} atoms C.
- You need two conversions to solve this problem. Your first task is to get to moles of C. To do this, use Avogadro's number. The 6.02×10^{23} atoms C needs to be on the bottom to cancel atoms and get you to moles.
- Your second task is to take the moles of C and convert to grams. To do this, you need to use the molar mass of C. This comes directly from the periodic table. I'm using one with few decimal points, so 12.0 g C goes on top and 1 mole C goes on the bottom of the conversion factor.
- Solve by multiplying values that are on top and dividing by numbers on the bottom.
- Start with 9.47×10^{25} atoms C divide by 6.02×10^{23} atoms C and multiply by 12.0 g C.
- The answer is 1887.707641 but to 3 SF that becomes 1890 g C.

Name: ________________________________ Date: _______________ Period: ______

Worksheet 5: FM, MM, Mole Conversions Self-quiz

Directions: For formula mass and molar mass problems, you can use short form. For convenience, just report all numbers for formula mass and molar mass problems. Use the conversion factor method to solve each of the problems where applicable.

1. What is the formula mass of nickel (III) bicarbonate?

2. What is the molar mass of cesium phosphate?

3. How many moles of titanium are equal to 7.39×10^{15} atoms of titanium?

4. How many formula units of sodium nitride are equal to 6.58 g of sodium nitride?

5. How many g of silicon dioxide are contained in 4.0 moles of silicon dioxide?

6. What is the mass of 8.2×10^{33} formula units of gold (III) nitrate?

Name: ________________________________ Date: _______________ Period: ______

KEY for Worksheet 5: FM, MM, Mole Conversions Self-quiz

Directions: For formula mass and molar mass problems, you can use short form. For convenience, just report all numbers for formula mass and molar mass problems. Use the conversion factor method to solve each of the problems where applicable.

1. What is the formula mass of nickel (III) bicarbonate?

Formula: $Ni(HCO_3)_3$

1 Ni x 58.7 amu = 58.7 amu
3 H x 1.0 amu = 3.0 amu
3 C x 12.0 amu = 36.0 amu
9 O x 16.0 amu = +144.0 amu
241.7 amu

2. What is the molar mass of cesium phosphate?

Formula: Cs_3PO_4

3 Cs x 132.9 g = 398.7 g Cs
1 P x 31.0 g = 31.0 g P
4 O x 16.0 g = + 64.0 g O
493.7 g/mol Cs_3PO_4

3. How many moles of titanium are equal to 7.39×10^{15} atoms of titanium?

$$7.39 \times 10^{15} \text{ atoms Ti} \times \frac{1\ mol\ Ti}{6.02 \times 10^{23} atoms\ Ti} = 1.23 \times 10^{-8} \text{ mol Ti}$$

4. How many formula units of sodium nitride are equal to 6.58 g of sodium nitride?

Formula: Na_3N

$$6.58 \text{ g } Na_3N \times \frac{1\ mol\ \text{Na3N}}{83\ g\ \text{Na3N}} \times \frac{6.02 \times 10^{23} form\ u\ \text{Na3N}}{1\ mol\ \text{Na3N}} = 4.77 \times 10^{22} \text{ form u } Na_3N$$

5. How many g of silicon dioxide are contained in 4.0 moles of silicon dioxide?

Formula: SiO_2

$$4.0 \text{ mol } SiO_2 \times \frac{60.1\ g\ \text{SiO2}}{1\ mol\ \text{SiO2}} = 240 \text{ g } SiO_2$$

6. What is the mass of 8.2×10^{33} formula units of gold (III) nitrate?

Formula: $Au(NO_3)_3$

$$8.2 \times 10^{33} \text{ form u } Au(NO_3)_3 \times \frac{1\ mol\ \text{Au(NO3)3}}{6.02 \times 10^{23} form\ u\ \text{Au(NO3)3}} \times \frac{383\ g\ \text{Au(NO3)3}}{1\ mol\ \text{Au(NO3)3}} = 5.2 \times 10^{12} \text{ g } Au(NO_3)_3$$

Percent Composition:

Steps to solving percent composition (% by mass) problems:

Step #1: Write down the chemical formula for the compound. (Sometimes, you may be given this information.)
Step #2: Find the mass of each element in the compound.
Step #3: Add the answers to get the total mass of the compound.
Step #4: For each element, take the mass of that element and divide it by the total mass of the compound.
Step #5: Multiply by 100 to turn this number into a percent.
Step #6: Check your answer by adding the percent values together. They should be very close to 100%.

Example: What is the percent composition of aluminum sulfate?
Step #1: chemical formula: $Al_2(SO_4)_3$
Step #2: Find the mass of each element in the compound.
2 Al x 27 g = 54 g
3 S x 32.1 g = 96.3 g
12 O x 16.0 g = 192 g

Tip: Leave extra room next to oxygen because; it's tempting for some students to think it's a zero.
Note: Your periodic table may have different values; I'm using one that only has one decimal place.

Step #3: Add the answers to get the total mass of the compound.
2 Al x 27 g = 54 g
3 S x 32.1 g = 96.3 g
12 O x 16.0 g = <u>192 g</u>
342.3 g/mol
Step #4: For each element, take the mass of that element and divide it by the total mass of the compound.

% Al: $\frac{54\ g}{342.3\ g}$ = .1578
% S: $\frac{96.3\ g}{342.3\ g}$ = .2813
%O: $\frac{192\ g}{342.3\ g}$ = .5609

Note: I dropped the per mol part on the part where I'm calculating percent. By the end, the mass units should cancel, giving me only the percent sign as a unit.

Step #5: Multiply by 100 to turn this number into a percent.

% Al: $\frac{54\ g}{342.3\ g}$ = .1578 x 100 = 15.78%

% S: $\frac{96.3\ g}{342.3\ g}$ = .2813 x 100 = 28.13%

%O: $\frac{192\ g}{342.3\ g}$ = .5609 x 100 = 56.09%

Step #6: Check your answer by adding the percent values together. They should be very close to 100.

15.78% + 28.13% + 56.09% = 100%

Let's go the other way …

If you know the percent composition and the mass of a compound, you can calculate the contribution of a particular element to that compound.

Example: What mass of oxygen is contained in a 14.5 g sample of $Al_2(SO_4)_3$?

Answer: 14.5 g $Al_2(SO_4)_3$ x .5609 = 8.13305 g

To three sig figs, the answer is 8.13 g

Explanation: I'm using this compound since I just figured out the percent of oxygen in the compound. If 56.09% of the compound is oxygen, then it follows that oxygen should make up 56.09% of the mass in any sample. Therefore, I solve this by turning the percent into a decimal by dividing by 100 (undoing the percent) and multiply by the mass of the sample.

Name: ________________________________ Date: _______________ Period: ______

Worksheet 6: Percent Composition

Directions: For simplicity, please keep all decimals as you work through each problem (don't round), and report the % answers to two decimal places. Don't forget units.

Note: If a chemical formula has parentheses, I will use brackets to offset it from the rest of the question. Remember to distribute the subscript to anything inside the parentheses.

1. What is the percent composition of table salt ($NaCl$)?

2. What is the percent composition of aluminum hydroxide [$Al(OH)_3$]?

3. What is the percent composition of green tea ($C_{22}H_{18}O_{11}$)?

4. What is the percent composition of sodium bicarbonate?
Formula:

5. What is the percent composition of iron (III) nitrate?
Formula:

Name: ________________________________ Date: _______________ Period: ______

KEY for Worksheet 6: Percent Composition

Directions: For simplicity, please keep all decimals as you work through each problem (don't round), and report the % answers to two decimal places. Don't forget units.

Note: If a chemical formula has parentheses, I will use brackets to offset it from the rest of the question. Remember to distribute the subscript to anything inside the parentheses.

1. What is the percent composition of table salt (NaCl)?

1 Na x 23.0 g = 23.0 g
1 Cl x 35.5 g = +35.5 g
58.5 g/mol NaCl

% Na: $\frac{23.0\ g\ Na}{58.5\frac{g}{mol}NaCl}$ x 100 = 39.32% Na

% Cl: $\frac{35.5\ g\ Cl}{58.5\frac{g}{mol}NaCl}$ x 100 = 60.68% Cl

39.32% + 60.68% = 100%

Explanation: Start by figuring out the molar mass of NaCl. The chemical formula consists of 1 Na with a mass of 23.0 g and 1 Cl with a mass of 35.5 g. Together, this gives you the mass of NaCl, which is 58.5 g/mol of the compound. To find the percent of Na in the compound, take the mass of Na (23.0 g) and divide by the molar mass value and multiply by 100. The percent of Na is 39.32%. Repeat this process for the Cl. The mass of Cl was 35.5 g. When this is divided by 58.5 g/mol NaCl and multiplied by 100, the percent is 60.68%. To check, you should add the percents. They should come very close to 100.

A note about units and labels. I will show you the long version in this first one, but then, I'll just shorten the percent portion to grams only as the unit. I will also stop putting the chemical label after the percent because I have put it before doing the calculation.

2. What is the percent composition of aluminum hydroxide [$Al(OH)_3$]?

1 Al x 27.0 g = 27.0 g
3 O x 16.0 g = 48.0 g
3 H x 1.0 g = +3.0 g
78.0 g/mol $Al(OH)_3$

% Al: $\frac{27.0\ g}{78.0\ g}$ x 100 = 34.62%

% O: $\frac{48.0\ g}{78.0\ g}$ x 100 = 61.54%

% H: $\frac{3.0\ g}{78.0\ g}$ x 100 = 3.85%

34.62% + 64.54% + 3.85% = 100.01%

Explanation: Multiply the number of each chemical in the compound by its molar mass. Add the masses to get the molar mass of the entire compound. Divide each chemical mass contribution by the compound's molar mass and multiply by 100. Check the answer by adding the percents. They should come near 100.

3. What is the percent composition of green tea ($C_{22}H_{18}O_{11}$)?

22 C x 12.0 g = 264.0 g
18 H x 1.0 g = 18.0 g
11 O x 16.0 g =+176.0 g
458.0 g/mol $C_{22}H_{18}O_{11}$

% C: $\frac{264.0\ g}{458.0\ g}$ x 100 = 57.64%

% H: $\frac{18.0\ g}{458.0\ g}$ x 100 = 3.93%

% O: $\frac{176.0\ g}{458.0\ g}$ x 100 = 38.43%

57.64% + 3.93% + 38.43% = 100%

Explanation:

- Multiply the number of each chemical in the compound by its molar mass. 22 carbons have a mass of 264.0 g, 18 H have a mass of 18.0 g, and 11 O have a mass of 176.0 g.
- Add the masses to get the molar mass of the entire compound. 264.0 g + 18.0 g + 176.0 g = 458.0 g/mol $C_{22}H_{18}O_{11}$
- Divide each chemical mass contribution by the compound's molar mass and multiply by 100.
  - The mass of carbon is 264.0 g. This divided by 458.0 g and then multiplied by 100 gives me 57.64%.
  - The mass of hydrogen is 18.0 g. This divided by 458.0 g and then multiplied by 100 gives me 3.93%.
  - The mass of oxygen is 176.0 g. This divided by 458.0 g and then multiplied by 100 gives me 38.43%.

- Check the answer by adding the percents. They should come near 100. 57.64% + 3.93% + 38.43% = 100%

4. What is the percent composition of sodium bicarbonate?
Formula: $NaHCO_3$

1 Na x 23.0 g = 23.0 g
1 H x 1.0 g = 1.0 g
1 C x 12.0 g = 12.0 g
3 O x 16.0 g = +48.0 g
84.0 g/mol $NaHCO_3$

% Na: $\frac{23.0\ g}{84.0\ g}$ x 100 = 27.38%

% H: $\frac{1.0\ g}{84.0\ g}$ x 100 = 1.19%

% C: $\frac{12.0\ g}{84.0\ g}$ x 100 = 14.29%

% O: $\frac{48.0\ g}{84.0\ g}$ x 100 = 57.14%

27.38% + 1.19% + 14.29% + 57.14% = 100%

Note: I'm not going to do word explanations for 4 and 5 because I think the math is much clearer without the words.

5. What is the percent composition of iron (III) nitrate?
Formula: $Fe(NO_3)_3$

1 Fe x 55.8 g = 55.8 g
3 N x 14.0 g = 42.0 g
9 O x 16.0 g = +144.0 g
241.8 g/mol $Fe(NO_3)_3$

% Fe: $\frac{55.8\ g}{241.8\ g}$ x 100 = 23.08%

% N: $\frac{42.0\ g}{241.8\ g}$ x 100 = 17.37%

% O: $\frac{144.0\ g}{241.8\ g}$ x 100 = 59.55%

23.08% + 17.37% + 59.55% = 100%

Percent of Water in a Hydrate:

What is a hydrate?

A hydrate is a chemical compound that contains water as part of its structure.

Sample Hydrates:

- Sodium sulfate decahydrate – $Na_2SO_4 \cdot 10\ H_2O$
- Zinc nitrate hexahydrate – $Zn(NO_3)_2 \cdot 6\ H_2O$

Formatting Note: There is a dot between the ionic formula portion of the chemical formula and the hydrate portion.

Naming Note: The name of a hydrate is the name of the ionic formula then the word hydrate preceded by a prefix that indicates the number of water molecules.

Quick review of prefixes 1-10:

- Mono = 1
- Di = 2
- Tri = 3
- Tetra = 4
- Penta = 5
- Hexa = 6
- Hepta = 7
- Octa = 8
- Nona = 9
- Deca = 10

Steps to solving percent water in a hydrate problems:

Note: It's exactly like percent by mass. The only difference is that you're only interested in the portion that is water.

Step #1: Write down the chemical formula for the compound. (Sometimes, you may be given this information.)
Step #2: Calculate the mass of the ionic salt portion of the formula.
Step #3: Calculate the mass of the water portion of the formula.
Step #4: Add the answers to get the total mass of the compound. Please note that this includes the water portion.
Step #5: Take the water's mass and divide it by the total mass of the compound, then multiply by 100 to turn it into a percent. For convenience, I have my students round to 2 decimal places.
Step #6: Check your answer by figuring out the percent of the non-water portion and add the two answers together. They should be very close to 100%.

Example: What is the percent composition of Sodium sulfate decahydrate?

Step #1: Formula: $Na_2SO_4 \bullet 10\ H_2O$

Step #2: Calculate the mass of the ionic salt portion of the formula.

2 Na x 23.0 g = 46.0 g
1 S x 32.1 g = 32.1 g
4 O x 16.0 g = +64.0 g
142.1 g/mol Na_2SO_4

Step #3: Calculate the mass of the water portion of the formula.

20 H x 1.0 g = 20.0 g
10 O x 16.0 g = +160.0 g
180.0 g/mol H_2O

Step #4: Find the total mass for the compound: 142.1 g/mol Na_2SO_4 + 180.0 g/mol H_2O = 322.1 g/mol $Na_2SO_4 \bullet 10\ H_2O$

Step #5: Take the water's mass and divide it by the total mass of the compound, then multiply by 100 to turn it into a percent. For convenience, I have my students round to 2 decimal places.

$$\frac{180.0\ \text{g/mol H2O}}{322.1\ \text{g/mol Na2SO4}\cdot 10\ \text{H2O}} \times 100 = 55.88\%$$

Step #6: Check your answer by figuring out the percent of the non-water portion and add the two answers together. They should be very close to 100%.

Check: $$\frac{142.1\ \text{g/mol Na2SO4}}{322.1\ \text{g/mol Na2SO4}\cdot 10\ \text{H2O}} \times 100 = 44.12\%$$

44.12% + 55.88% = 100%

Name: ______________________________ Date: ______________ Period: ______

Worksheet 7: Percent of Water in a Hydrate

Directions: For simplicity, please keep all decimals as you work through each problem (don't round), and report the % answers to two decimal places.

1. What is the percent of water in iron (III) nitrate nonahydrate [$Fe(NO_3)_3 \cdot 9H_2O$]?

2. What is the percent of water in manganese (II) bromide tetrahydrate [$MnBr_2 \cdot 4H_2O$]?

3. What is the percent of water in tin (IV) chloride pentahydrate?
Formula:

Name: ______________________________ Date: ______________ Period: _____

KEY for Worksheet 7: Percent of Water in a Hydrate

Directions: For simplicity, please keep all decimals as you work through each problem (don't round), and report the % answers to two decimal places.

1. What is the percent of water in iron (III) nitrate nonahydrate [$Fe(NO_3)_3 \cdot 9H_2O$]?

1 Fe x 55.8 g = 55.8 g
3 N x 14.0 g = 42.0 g
9 O x 16.0 g =+144.0 g
241.8 g/mol $Fe(NO_3)_3$

18 H x 1.0 g = 18.0 g
9 O x 16.0 g =+144.0 g
162 g/mol H_2O

Total mass of the compound: 241.8 g/mol + 162 g/mol = 403.8 g/mol $Fe(NO_3)_3 \cdot 9H_2O$

Percent of water in the hydrate: $\frac{162 \frac{g}{mol} \text{ H2O}}{403.8 \frac{g}{mol} \text{ Fe(NO3)3·9H2O}} \times 100 = 40.12\%$

Explanation: First obtain the mass of the anhydrous salt (the part without water; 241.8 g/mol) and the mass of the water (162 g/mol). Then add these values to obtain the mass of the compound (403.8 g/mol $Fe(NO_3)_3 \cdot 9H_2O$). Take the mass of the water (241.8 g/mol) and divide by the mass of the whole compound (403.8 g/mol $Fe(NO_3)_3 \cdot 9H_2O$) and multiply by 100. The percent will be 40.12% if rounded to two decimal places.

If you want to check your answer, take the mass of the anhydrous salt (241.8 g/mol $Fe(NO_3)_3$) and divide by the mass of the compound (403.8 g/mol $Fe(NO_3)_3 \cdot 9H_2O$) and multiply by 100. These two values, the percent obtained here (59.88%) and the percent obtained by the water only (40.12%) should be very close to 100%.

2. What is the percent of water in manganese (II) bromide tetrahydrate [$MnBr_2 \cdot 4H_2O$]?

1 Mn x 54.9 g = 54.9 g
2 Br x 79.9 g =+159.8 g
214.7 g/mol $MnBr_2$

8 H x 1.0 g = 8.0 g
4 x 16.0 g =+64.0 g
72.0 g/mol H_2O

Total mass of the compound: 214.7 g/mol + 72.0 g/mol = 286.7 g/mol $MnBr_2 \cdot 4H_2O$

Percent of water in the hydrate: $\frac{72.0 \frac{g}{mol} H2O}{286.7 \frac{g}{mol} \text{MnBr2·4H2O}}$ x 100 = 25.11%

Explanation: First obtain the mass of the anhydrous salt (the part without water; 214.7 g/mol $MnBr_2$) and the mass of the water (72.0 g/mol H_2O). Then add these values to obtain the mass of the compound (286.7 g/mol $MnBr_2 \cdot 4H_2O$). Take the mass of the water (72.0 g/mol H_2O) and divide by the mass of the whole compound (286.7 g/mol $MnBr_2 \cdot 4H_2O$) and multiply by 100. The percent will be 25.11% if rounded to two decimal places.

If you want to check your answer, take the mass of the anhydrous salt (214.7 g/mol $MnBr_2$) and divide by the mass of the compound (286.7 g/mol $MnBr_2 \cdot 4H_2O$). and multiply by 100. These two values, the percent obtained here (74.89%) and the percent obtained by the water only (25.11%) should be very close to 100%.

3. What is the percent of water in tin (IV) chloride pentahydrate?

Formula: $SnCl_4 \cdot 5H_2O$

1 Sn x 118.7 g = 118.7 g
4 Cl x 35.5 g = +142.0 g
260.7 g/mol $SnCl_4$

10 H x 1.0 g = 10.0 g
5 O x 16.0 g = +80.0 g
90.0 g/mol H_2O

Total mass of the compound: 260.7 g/mol + 90.0 g/mol = 350.7 g/mol $SnCl_4 \cdot 5H_2O$

Percent of water in the hydrate: $\frac{90.0 \frac{g}{mol} H2O}{350.7 \frac{g}{mol} \text{SnCl4·5H2O}}$ x 100 = 25.66%

Explanation: First obtain the mass of the anhydrous salt (the part without water; 260.7 g/mol $SnCl_4$) and the mass of the water (90.0 g/mol H_2O). Then add these values to obtain the mass of the compound (350.7 g/mol $SnCl_4 \cdot 5H_2O$). Take the mass of the water (90 g/mol H_2O) and divide by the mass of the whole compound (350.7 g/mol $SnCl_4 \cdot 5H_2O$) and multiply by 100. The percent will be 25.66% if rounded to two decimal places.

If you want to check your answer, take the mass of the anhydrous salt (260.7 g/mol $SnCl_4$) and divide by the mass of the compound (350.7 g/mol $SnCl_4 \cdot 5H_2O$) and multiply by 100. These two values, the percent obtained here (74.34%) and the percent obtained by the water only (25.66%) should be very close to 100%.

Name: ______________________________ Date: ______________ Period: ______

Worksheet 8: Percent Calculations Self-quiz

Directions: For simplicity, please keep all decimals as you work through each problem (don't round) and report the % answers to two decimal places.

1. What is the percent composition of strontium oxide?
Formula:

2. What is the percent composition of coffee ($C_{25}H_{28}N_6O_7$)?

3. What is the percent of water in iron (II) sulfate heptahydrate?

Name: ______________________________ Date: ______________ Period: ______

KEY for Worksheet 8: Percent Calculations Self-Quiz

Directions: For simplicity, please keep all decimals as you work through each problem (don't round), and report the % answers to two decimal places.

1. What is the percent composition of strontium oxide?
Formula: SrO

1 Sr x 87.6 g = 87.6 g
1 O x 16.0 g = +16.0 g
103.6 g/mol SrO

% Sr: $\frac{87.6\ g}{103.6\ g}$ x 100 = 84.56%

% O: $\frac{16.0\ g}{103.6\ g}$ x 100 = 15.44%

Check: 84.56% + 15.44% = 100%

2. What is the percent composition of coffee ($C_{25}H_{28}N_6O_7$)?

25 C x 12.0 g = 300.0 g
28 H x 1.0 g = 28.0 g
6 N x 14.0 g = 84.0 g
7 O x 16.0 g = +112.0 g
524 g/mol $C_{25}H_{28}N_6O_7$

% C: $\frac{300.0\ g}{524\ g}$ x 100 = 57.25%

% H: $\frac{28.0\ g}{524\ g}$ x 100 = 5.34%

% N: $\frac{84.0\ g}{524\ g}$ x 100 = 16.03%

% O: $\frac{112.0\ g}{524\ g}$ x 100 = 21.37%

57.25 % + 5.35% + 16.03% + 21.37% = 100%

3. What is the percent of water in iron (II) sulfate heptahydrate?
Formula: $FeSO_4 \cdot 7H_2O$

1 Fe x 55.8 g = 55.8 g
1 S x 32.1 g = 32.1 g
4 O x 16.0 g = <u>+64.0 g</u>
151.9 g/mol $FeSO_4$

14 H x 1.0 g = 14.0 g
7 O x 16.0 g = <u>+112.0 g</u>
126 g/mol H_2O

Total mass of the compound: 151.9 g/mol + 126 g/mol = 277.9 g/mol $FeSO_4 \cdot 7H_2O$

Percent of water in the hydrate: $\frac{126 \frac{g}{mol} H2O}{277.9 \frac{g}{mol} \text{FeSO4·7H2O}}$ x 100 = 45.34%

Empirical and Molecular Formulas:

Empirical formula: simplest form of a chemical formula; i.e. CH_2
Molecular formula: actual chemical formula for a compound; i.e. C_2H_4

Note: These formulas could be the same.

Steps to finding empirical formulas:

Step #1: Find the # of moles for each element by doing a molar mass conversion to moles.

Note: If you're given percent values, then assume a 100 g sample. Therefore, your # of g = the %.

Step #2: Find the mole ratio of the elements to each other. If the numbers come out as a decimal, multiply them by a value that will make everything a whole number.

Step #3: Write the whole numbers from the ratio as the subscripts for each of the elements.

Example:

Find the empirical formula of the chemical compound given that it contains 83.33% C and 16.67% H.

Step #1: Convert to moles.

$$83.33 \text{ g C} \times \frac{1\,mol\,C}{12\,g\,C} = 6.94 \text{ mol C}$$

$$16.67 \text{ g H} \times \frac{1\,mol\,H}{1.0\,g\,H} = 16.67 \text{ mol H}$$

Step #2: Find the mole ratio of the elements to each other. If the numbers come out as a decimal, multiply them by a value that will make everything a whole number.
Mole Ratio:
C : H

$$\frac{6.94}{6.94} : \frac{16.67}{6.94}$$

(1 : 2.4) x 5 = 5 : 12

Step #3: Write the whole numbers from the ratio as the subscripts for each of the elements.

Empirical formula: C_5H_{12}

Steps to calculating molecular formulas:

Step #1: Calculate the empirical formula.

Step #2: Calculate the formula mass/molar mass for the empirical formula.

Step #3: Divide the molecular formula mass/molar mass by the empirical mass to get a small whole number.

Step #4: Multiply the empirical formula by that small whole number value to get the molecular formula. I.e. If your empirical formula is CH_2 and your small whole number is 3, your molecular formula will be 3 x CH_2 = C_3H_6.

Example:

A compound has a molar mass of 114 g/mol and a percent composition of 84.21% C and 15.79% H. Calculate the empirical and molecular formulas.

Step #1: Assume we worked through the steps above and got the empirical formula of: C_4H_9

Step #2: Calculate the molar mass for the empirical formula: (4 C x 12 g) + (9 H x 1.0 g) = 57 g/mol

Step #3: Divide the molecular mass by the empirical mass: $\frac{114\frac{g}{mol}}{57\frac{g}{mol}} = 2$

Step #4: Multiply the empirical formula by the whole number to get the molecular formula: C_8H_{18}

Name: ______________________________ Date: ______________ Period: _____

Worksheet 9: Calculating Empirical Formulas

Directions: When converting to moles, please save at least 2-3 decimal places.
Note: If your whole number ratio turns out to be a decimal, multiply it by a small whole number until you get a whole number.

1. What is the empirical formula for a compound containing 15.93% B and 84.07% F?
Step 1: Convert grams into moles

Step 2: Set up a mole ratio

Step 3: Turn the ratio into a chemical formula

2. What is the empirical formula for a compound containing 40% C, 6.67% H, and 53.33% O?
Step 1: Convert grams into moles

Step 2: Set up a mole ratio

Step 3: Turn the ratio into a chemical formula

3. What is the empirical formula for a compound containing 82.76% C and 17.24% H?
Step 1: Convert grams into moles

Step 2: Set up a mole ratio

Step 3: Turn the ratio into a chemical formula

Name: ______________________________ Date: ______________ Period: _____

KEY for Worksheet 9: Calculating Empirical Formulas

Directions: When converting to moles, please save at least 2-3 decimal places.
Note: If your whole number ratio turns out to be a decimal, multiply it by a small whole number until you get a whole number.

1. What is the empirical formula for a compound containing 15.93% B and 84.07% F?

Step 1: Convert grams into moles

15.93 g B x $\frac{1\ mol\ B}{10.8\ g\ B}$ = 1.475 mol B

84.07 g F x $\frac{1\ mol\ F}{19.0\ g\ F}$ = 4.42 mol F

Explanation: Grams are converted to moles by setting up a conversion factor that goes from grams on the bottom to moles on the top. The number of grams is the mass of the compound and can be found on the periodic table of elements. Do this for each element.

Step 2: Set up a mole ratio

Moles of B: Moles of F

$$\frac{1.475}{1.475} : \frac{4.42}{1.475}$$

$$1 : 3$$

Explanation: I usually let my students skip writing moles everywhere as long as they're clear about setting up a mole ratio. You get a whole number ratio by first writing the number of moles of each chemical under their respective symbols. Then, divide through by the smaller or smallest number of moles. In this case, that was the 1.475 moles of B.

Step 3: Turn the ratio into a chemical formula

Chemical formula: BF_3

Explanation: The ratio can be turned into a chemical formula by writing the numbers that are part of the ratio as the subscripts next to the chemical symbols. Note that a subscript of 1 is not used. If there is no number, the assumption is 1.

2. What is the empirical formula for a compound containing 40% C, 6.67% H, and 53.33% O?

Step 1: Convert grams into moles

40.0 g C x $\frac{1\ mol\ C}{12.0\ g\ C}$ = 3.33 mol C

6.67 g H x $\frac{1\ mol\ H}{1\ g\ H}$ = 6.67 mol H

53.33 g O x $\frac{1\ mol\ O}{16.0\ g\ O}$ = 3.33 mol O

Step 2: Set up a mole ratio

Moles of C: Moles of H: Moles of O

$$\frac{3.33}{3.33} : \frac{6.67}{3.33} : \frac{3.33}{3.33}$$

$$1 : 2 : 1$$

Step 3: Turn the ratio into a chemical formula

Chemical Formula: CH_2O

3. What is the empirical formula for a compound containing 82.76% C and 17.24% H?

Step 1: Convert grams into moles

$$82.76 \text{ g C x } \frac{1 \, mol \, C}{12 \, g \, C} = 6.90 \text{ mol C}$$

$$17.24 \text{ g H x } \frac{1 \, mol \, H}{1 \, g \, H} = 17.24 \text{ mol H}$$

Note: I know it's annoying to set up, especially when the molar mass is 1 g/mol. However, not every periodic table will be as simplistic as the one I'm currently using.

Step 2: Set up a mole ratio

Moles of C: Moles of H

$$\frac{6.90}{6.90} : \frac{17.24}{6.90}$$

$$1 : 2.5$$

2 x (1:2.5) = 2:5

Note: Dividing the moles did not give us a whole number ratio, but 2.5 can become a whole number once multiplied by 2. Both numbers have to be treated the same, which is how we got 2:5.

Step 3: Turn the ratio into a chemical formula

Chemical formula: C_2H_5

Name: ________________________________ Date: _______________ Period: ______

Worksheet 10: Calculating Molecular Formulas

Directions: When converting to moles, please save at least 2-3 decimal places.

Note: Steps 1-3 are that of finding the empirical formula.
Second Note: If the small whole number in step 5 is a 1, that means that the empirical and molecular formulas are identical.
Third Note: These are longer problems. If you require more room than is provided, I recommend working on a sheet of printer paper.

1. What is the molecular formula for a compound with a mass of 162 amu that contains 74.07% C, 8.64% H, and 17.28% N?
Step 1: Convert grams into moles

Step 2: Set up a mole ratio

Step 3: Turn the ratio into a chemical formula
Step 4: Find the mass of the empirical formula
Step 5: Divide the molecule's mass by the empirical formula's mass. This will result in a small whole number.
Step 6: Multiply the empirical formula by this small whole number to obtain the molecule's formula.

2. What is the molecular formula for a compound with a formula mass of 284 amu that contains 43.66% P and 56.34% O?
Step 1: Convert grams into moles

Step 2: Set up a mole ratio

Step 3: Turn the ratio into a chemical formula
Step 4: Find the mass of the empirical formula
Step 5: Divide the molecule's mass by the empirical formula's mass. This will result in a small whole number.
Step 6: Multiply the empirical formula by this small whole number to obtain the molecule's formula.

Name: ______________________________ Date: ______________ Period: _____

KEY for Worksheet 10: Calculating Molecular Formulas

Directions: When converting to moles, please save at least 2-3 decimal places.

Note: Steps 1-3 are that of finding the empirical formula.
Second Note: If the small whole number in step 5 is a 1, that means that the empirical and molecular formulas are identical.
Third Note: These are longer problems. If you require more room than is provided, I recommend working on a sheet of printer paper.

1. What is the molecular formula for a compound with a mass of 162 amu that contains 74.07% C, 8.64% H, and 17.28% N?

Step 1: Convert grams into moles

74.07 g C x $\frac{1\ mol\ C}{12.0\ g\ C}$ = 6.17 mol C

8.64 g H x $\frac{1\ mol\ H}{1\ g\ H}$ = 8.64 mol H

17.28 g N x $\frac{1\ mol\ N}{14.0\ g\ N}$ = 1.23 mol N

Explanation: When given percents, the assumption is that there is a 100 g sample. Each of these grams are converted to moles by dividing the number given by the molar mass. When the conversion factor is set up, the mass is on the bottom and the moles is on top.

Step 2: Set up a mole ratio

Moles of C: Moles of H: Moles of N

$\frac{6.17}{1.23} : \frac{8.64}{1.23} : \frac{1.23}{1.23}$

5 : 7 : 1

Explanation: The number of moles of each element in the formula is turned into a whole number ratio by dividing each number of moles by the smallest of these. We got whole numbers, so there is nothing more to be done for this step.

Step 3: Turn the ratio into a chemical formula: C_5H_7N

Explanation: The numbers from the whole number ratio of elements are turned into the chemical formula by placing the numbers as a subscript next to their respective element symbols. The N does not get a subscript here because it would have been 1.

Step 4: Find the mass of the empirical formula

5 C x 12.0 amu = 60.0 amu

7 H x 1.0 amu = 7.0 amu

1 N x 14.0 amu = 14.0 amu

Mass of empirical formula: 60.0 amu + 7.0 amu + 14.0 amu = 81.0 amu

Note the switch to amu. It's numerically the same as molar mass, but the molecule's mass was given to us in amu, so we should match the units.

Explanation: This step is solved exactly as you would if the question was "What is the formula mass of C_5H_7N?". Each subscript is multiplied by the mass of that element. Then, these values are added.

Step 5: Divide the molecule's mass by the empirical formula's mass. This will result in a small whole number.

$$\frac{mass\ of\ the\ molecule}{mass\ of\ the\ empirical\ formula} = \frac{162\ amu}{81.0\ \text{amu}} = 2$$

Explanation: We were given the molecule's mass early in the problem. Sometimes, it's hard to remember that. You should always get a small whole number because the molecule is some multiple of the empirical formula.

Step 6: Multiply the empirical formula by this small whole number to obtain the molecule's formula. 2x C_5H_7N = $C_{10}H_{14}N_2$

Note: If the small whole number is a 1, that means that the empirical and molecular formulas are identical.

Explanation: The small whole number was 2. The empirical formula was C_5H_7N. When each of the subscripts is doubled, we get the final answer, which is $C_{10}H_{14}N_2$.

2. What is the molecular formula for a compound with a formula mass of 284 amu that contains 43.66% P and 56.34% O?

Step 1: Convert grams into moles

$43.66\ \text{g P} \times \frac{1\ mol\ P}{31.0\ g\ P} = 1.408\ \text{mol P}$

$56.34\ \text{g O} \times \frac{1\ mol\ O}{16.0\ g\ O} = 3.521\ \text{mol O}$

Note: I have chosen 3 decimal places this time because I didn't want to round the 8. It should still work if you go to 2 decimals.

Step 2: Set up a mole ratio

Moles of P: Moles of O

$\frac{1.408}{1.408} : \frac{3.521}{1.408}$

1 : 2.50

2 x (1:2.5) = 2:5

Step 3: Turn the ratio into a chemical formula: P_2O_5

Step 4: Find the mass of the empirical formula

2 P x 31.0 amu = 62.0 amu

5 O x 16.0 amu = 80.0 amu

Mass of the empirical formula: 62.0 amu + 80.0 amu = 142 amu

Step 5: Divide the molecule's mass by the empirical formula's mass. This will result in a small whole number.

$$\frac{mass\ of\ the\ molecule}{mass\ of\ the\ empirical\ formula} = \frac{284\ \text{amu}}{142\ \text{amu}} = 2$$

Step 6: Multiply the empirical formula by this small whole number to obtain the molecule's formula.

2 x P_2O_5 = P_4O_{10}

Calculating the Number of Water Molecules in a Hydrate:

What is a hydrate?
In chemistry, it's a compound that has water as part of its formula.

Anhydrous crystal: a hydrate without the water; this is what you have left after burning off the water; that's a fun lab

Steps to finding moles of water in a hydrate: (this becomes the # of water molecules in that hydrate)

Step #1: Find the mass of the anhydrous (without water) crystal.

Step #2: Find the mass of the water.

Note: You will likely have to do to do some subtraction to find either the mass of the anhydrous or the mass of the water. It really depends on what information you are given.

Step #3: Find the moles of the anhydrous crystal and the moles of the water.

Step #4: Find the whole # ratio by dividing both mole amounts by the smallest amount of moles
Note: The crystal amount is always smaller.

Step #5: Write the formula with a coefficient in front of the water to represent the amount of molecules contained in the crystal structure. Put a dot in between the salt and the water.

Example: 4.78 g of anhydrous lithium perchlorate was dissolved in water and re-crystalized. Care was taken to isolate all the lithium perchlorate as its hydrate. The mass of the hydrated salt obtained was 7.21 g. What hydrate is it?
Step #1: Find mass of substance without water. 4.78 g

Step #2: Find mass of water. 7.21 g - 4.78 g = 2.43 g

Step #3: Find moles of salt and water.

$$4.78 \text{ g } LiClO_4 \times \frac{1 \text{ mol LiClO4}}{106.4 \text{ g LiClO4}} = 0.0449 \text{ mol}$$

$$2.43 \text{ g } H_2O \times \frac{1 \text{ mol H2O}}{18.0 \text{ g H2O}} = .135 \text{ mol } H_2O$$

Step #4: Whole # ratio:

$$LiClO_4: \frac{0.044929 \text{ mol}}{0.044929 \text{ mol}} = 1 \quad H_2O: \frac{0.13489 \text{ mol}}{0.044929 \text{ mol}} = 3$$

Step #5: Write the formula. $LiClO_4 \cdot 3H_2O$

You can also find the formula of the hydrate if you know the percent composition of the compound. Let's use the previous problem as an example.

$LiClO_4 \cdot 3H_2O$ has a mass of 160.4 g of which 54 g is water. The water is 33.67% of the compound. That means that the rest of the compound must be 66.33%. Then, you can find moles of each and set up a ratio.

Question: What is the formula for: $LiClO_4 \cdot n\ H_2O$

Step #1: Find mass of each. If you're given the percent of the water and the percent of the compound, then you can assume grams.

$$66.33\ g\ LiClO_4 \times \frac{1\ mol\ LiClO4}{106.4\ g\ LiClO4} = .623\ mol\ LiClO_4$$

$$33.67\ g\ H_2O \times \frac{1\ mol\ H2O}{18\ g\ H2O} = 1.87\ mol\ H_2O$$

Step #2: Set up a mole ratio.

$LiClO_4 : H_2O$

$\frac{.623}{.623} : \frac{1.87}{.623}$

1 : 3

Note: The part of the compound that's not water will always be the smaller number of moles.

Step #3: Write the empirical formula for the compound by plugging in the number of moles of water.

$LiClO_4 \cdot 3\ H_2O$

Name: ___________________________________ Date: _______________ Period: ______

Worksheet 11: Calculating the Number of Water Molecules in a Hydrate

Directions: When converting to moles, please save at least 2-3 decimal places.

Note: These steps should be very familiar to you. They are essentially how to find an empirical formula. The difference is that this ratio will always be a 1:something.
Reminder: Anhydrous means without water. It's referring to the salt only.
Second Note: n is the variable used for the number of moles of water in a hydrate. I have also seen x.

1. What is the value of n, the formula for $CaSO_4{\cdot}nH_2O$, and the name of the compound if a 7 g sample contains 1.46 g H_2O?
Step #1: Find mass of substance without water.

Step #2: Find mass of water.

Step #3: Find moles of salt and water.

Step #4: Whole # ratio:

Step #5: Write the formula (and answer any other questions posed).

2. If a 5.5 g sample of $Ni(NO_3)_2{\cdot}nH_2O$ was heated until the water was boiled off, the mass of the anhydrate left would be 3.45675 g $Ni(NO_3)_2$. What is the n? What is the formula? What is the name of the hydrate?
Step #1: Find mass of substance without water.

Step #2: Find mass of water.

Step #3: Find moles of salt and water.

Step #4: Whole # ratio:

Step #5: Write the formula (and answer any other questions posed).

Name: ______________________________ Date: ______________ Period: ______

KEY for Worksheet 11: Calculating the Number of Water Molecules in a Hydrate

Directions: When converting to moles, please save at least 2-3 decimal places.

Note: These steps should be very familiar to you. They are essentially how to find an empirical formula. The difference is that this ratio will always be a 1:something.
Reminder: Anhydrous means without water. It's referring to the salt only.
Second Note: n is the variable used for the number of moles of water in a hydrate. I have also seen x.

1. What is the value of n, the formula for $CaSO_4 \cdot nH_2O$, and the name of the compound if a 7 g sample contains 1.46 g H_2O?
Step #1: Find mass of substance without water. 7 g - 1.46 g = 5.54 g $CaSO_4$

Step #2: Find mass of water. This is given as 1.46 g H_2O.
Some years, to save space, I combine this into one step. Find the masses.

Step #3: Find moles of salt and water.

$$5.54 \text{ g } CaSO_4 \times \frac{1 \text{ mol CaSO4}}{136.2 \text{ g CaSO4}} = 0.04 \text{ mol } CaSO_4$$

$$1.46 \text{ g } H_2O \times \frac{1 \text{ mol H2O}}{18.0 \text{ g H2O}} = .08 \text{ mol } H_2O$$

Explanation: These are standard 1-step mole conversions. You start with the number of grams of each portion, the anhydrous salt ($CaSO_4$) and the water. Convert to moles by dividing by the mass portion of the molar mass. I chose not to keep more decimals because I saw that it would be a 1:2 ratio, but you can keep more. You might have to do some rounding in that case when you get to the ratio step.

Step #4: Whole # ratio:

$CaSO_4$: $\frac{0.04 \text{ mol}}{0.04 \text{ mol}} = 1$ H_2O: $\frac{0.08 \text{ mol}}{0.04 \text{ mol}} = 2$

Explanation: I changed the format of how I present the answer because I know there will only be two things in my mole ratio, but you don't have to. Here's the "normal" way of presenting the mole ratio.

Moles of $CaSO_4$: Moles of H_2O

$$\frac{0.04 \text{ mol}}{0.04 \text{ mol}} : \frac{0.08 \text{ mol}}{0.04 \text{ mol}}$$

1 : 2

Step #5: Write the formula (and answer any other questions posed).
n = 2
Formula: $CaSO_4 \cdot nH_2O$
Name: Calcium sulfate dihydrate

Behind the Scenes: To get the numbers for the problem, I had to find the percent of anhydrous salt and water in the hydrate. Then, I took the decimal percent (the percent divided by 100) and multiplied it by the number of grams I wanted in the sample.
$CaSO_4 \cdot 2H_2O$
136.2 g $CaSO_4$ + 36.0 g H_2O = 172.2 g/mol $CaSO_4 \cdot 2H_2O$

$$\% \ CaSO_4: \frac{136.2 \text{ g CaSO4}}{172.2 \text{ g/mol CaSO4·2H2O}} \times 100 = 79.09\%$$

$$\% \ H_2O: \frac{36.0 \text{ g H2O}}{172.2 \text{ g/mol CaSO4·2H2O}} \times 100 = 20.91\%$$

2. If a 5.5 g sample of $Ni(NO_3)_2 \cdot nH_2O$ was heated until the water was boiled off, the mass of the anhydrate left would be 3.45675 g $Ni(NO_3)_2$. What is the n? What is the formula? What is the name of the hydrate?
Step #1: Find mass of substance without water. This is given as 3.45675 g $Ni(NO_3)_2$.

Step #2: Find mass of water. 5.5 g hydrate - 3.45675 g $Ni(NO_3)_2$ = 2.04325 g H_2O
Some years, to save space, I combine this into one step. Find the masses.
You will always be given one of them and have to find the other.

Step #3: Find moles of salt and water.

$$3.45675 \text{ g } Ni(NO_3)_2 \times \frac{1 \text{ mol Ni(NO3)2}}{182.7 \text{ g Ni(NO3)2}} = 0.0189 \text{ mol } Ni(NO_3)_2$$

$$2.04325 \text{ g } H_2O \times \frac{1 \text{ mol H2O}}{18.0 \text{ g H2O}} = .1135 \text{ mol } H_2O$$

Explanation: These are standard 1-step mole conversions. You start with the number of grams of each portion, the anhydrous salt ($Ni(NO_3)_2$) and the water. Convert to moles by dividing by the mass portion of the molar mass. I chose to round to 4 decimals this time.

Step #4: Whole # ratio:

$$Ni(NO_3)_2: \frac{0.0189 \text{ mol}}{0.0189 \text{ mol}} = 1 \quad H_2O: \frac{.1135 \text{ mol}}{0.0189 \text{ mol}} = 6$$

Explanation: I changed the format of how I present the answer because I know there will only be two things in my mole ratio, but you don't have to. Here's the "normal" way of presenting the mole ratio.
Moles of $Ni(NO_3)_2$: Moles of H_2O

$$\frac{0.0189 \text{ mol}}{0.0189 \text{ mol}} : \frac{.1135 \text{ mol}}{0.0189 \text{ mol}}$$

$$1 : 6$$

Step #5: Write the formula (and answer any other questions posed).
n = 6
Formula: $Ni(NO_3)_2{\cdot}6H_2O$
Name: nickel (II) nitrate hexahydrate

Behind the Scenes Math:
Start with Nickel (II) nitrate hexahydrate $Ni(NO_3)_2{\cdot}6H_2O$
Get masses 182.7 g $Ni(NO_3)_2$ + 108 g H_2O = 290.7 g $Ni(NO_3)_2{\cdot}6H_2O$
Get the percent of the anhydrous salt and water:
% $Ni(NO_3)_2$: $\frac{182.7 \text{ g Ni(NO3)2}}{290.7 \text{ g Ni(NO3)2·6H2O}} \times 100 = 62.85\%$
% H_2O: $\frac{108 \text{ g H2O}}{290.7 \text{ g Ni(NO3)2·6H2O}} \times 100 = 37.15\%$
Randomly chosen number of grams to start with: 5.5 g $Ni(NO_3)_2{\cdot}nH_2O$
5.5 g of hydrate x decimal percent of anhydrate = 5.5 g x .6285 = 3.45675 g $Ni(NO_3)_2$

Name: ______________________________ Date: ______________ Period: ______

Worksheet 12: Empirical, Molecular, Hydrate Calculations Self-quiz

Directions: When converting to moles, please save at least 2-3 decimal places.

Reminder: Anhydrous means without water. It's referring to the salt only.
Note: There will only be a molecular problem, instead of empirical only and molecular because you need to find an empirical formula to get to the molecular formula.

1. What is the molecular formula for a compound with a formula mass of 78 amu that contains 92.31% C and 7.69% H?
Step 1: Convert grams into moles

Step 2: Set up a mole ratio

Step 3: Turn the ratio into a chemical formula:
Step 4: Find the mass of the empirical formula

Step 5: Divide the molecule's mass by the empirical formula's mass. This will result in a small whole number.

Step 6: Multiply the empirical formula by this small whole number to obtain the molecule's formula.

2. If a 6.84 g sample of $BeSO_3 \cdot nH_2O$ was heated until the water was boiled off, the mass of the anhydrate left would be 3.783 g $BeSO_3$. What is the value of n? What is the formula? What is the name of the hydrate?
Step #1: Find mass of substance without water.

Step #2: Find mass of water.

Step #3: Find moles of salt and water.

Step #4: Whole # ratio:

Step #5: Write the formula (and answer any other questions posed).

Name: ______________________________ Date: _______________ Period: ______

KEY for Worksheet 12: Empirical, Molecular, Hydrate Calculations Self-quiz

Directions: When converting to moles, please save at least 2-3 decimal places.

Reminder: Anhydrous means without water. It's referring to the salt only.
Note: There will only be a molecular problem, instead of empirical only and molecular because you need to find an empirical formula to get to the molecular formula.

1. What is the molecular formula for a compound with a formula mass of 78 amu that contains 92.31% C and 7.69% H?

Step 1: Convert grams into moles

$92.31 \text{ g C} \times \frac{1 \text{ mol } C}{12.0 \text{ g } C} = 7.69 \text{ mol C}$

$7.69 \text{ g H} \times \frac{1 \text{ mol } H}{1.0 \text{ g } H} = 7.69 \text{ mol H}$

Step 2: Set up a mole ratio

Moles of C: Moles of H

$\frac{7.69}{7.69} : \frac{7.69}{7.69}$

1 : 1

Step 3: Turn the ratio into a chemical formula: CH

Step 4: Find the mass of the empirical formula

1 C x 12.0 amu = 12.0 amu

1 H O x 1.0 amu = 1.0 amu

Mass of the empirical formula: 12.0 amu + 1.0 amu = 13 amu

Step 5: Divide the molecule's mass by the empirical formula's mass. This will result in a small whole number.

$\frac{\text{mass of the molecule}}{\text{mass of the empirical formula}} = \frac{78 \text{ amu}}{13 \text{ amu}} = 6$

Step 6: Multiply the empirical formula by this small whole number to obtain the molecule's formula.

6 x CH = C_6H_6

2. If a 6.84 g sample of $BeSO_3 \cdot nH_2O$ was heated until the water was boiled off, the mass of the anhydrate left would be 3.783 g $BeSO_3$. What is the value of n? What is the formula? What is the name of the hydrate?

Step #1: Find mass of substance without water. This is given as 3.783 g $BeSO_3$.

Step #2: Find mass of water. 6.84 g hydrate - 3.783 g $BeSO_3$ = 3.057 g H_2O

Step #3: Find moles of salt and water.

3.783 g $BeSO_3$ x $\frac{1\ mol\ BeSO3}{89.1\ g\ BeSO3}$ = .042 mol $BeSO_3$

3.057 g H_2O x $\frac{1\ mol\ H2O}{18.0\ g\ H2O}$ = .170 mol H_2O

Step #4: Whole # ratio:

Moles of $BeSO_3$: Moles of H_2O

$\frac{.042}{.042}$: $\frac{.170}{.042}$

1 : 4

Note: The answer comes to be 4.05. This isn't ideal, but it's close to 4. If you save more decimals, it will work out better.

Step #5: Write the formula (and answer any other questions posed).

n = 4

Formula: $BeSO_3{\cdot}4H_2O$

Name: Beryllium sulfite tetrahydrate

Behind the Scenes Math:

$BeSO_3{\cdot}4H_2O$

89.1 g $BeSO_3$ + 72.0 g H_2O = 161.1 g/mol $BeSO_3{\cdot}4H_2O$

% $BeSO_3$: $\frac{89.1\ g\ BeSO3}{161.1\ \frac{g}{mol}\ BeSO3{\cdot}4H2O}$ x 100 = 55.31%

% H_2O: $\frac{72.0\ g\ H2O}{161.1\ \frac{g}{mol}\ BeSO3{\cdot}4H2O}$ x 100 = 44.69%

Randomly selected mass 6.84 g $BeSO_3{\cdot}4H_2O$ x .5531 = 3.783 g $BeSO_3$

Name: ______________________________ Date: ______________ Period: ______

Worksheet 13: Practice Test 1

Directions: Show all work. Use the conversion factor method where appropriate. Go to significant figures for conversions. Formula masses and molar masses, keep all decimals. Round percentages to 2 decimals. Short form is fine.

1. What is the molar mass of titanium (II) acetate ($Ti(C_2H_3O_2)_2$)?

2. How many grams of strontium oxide are equal to 6.33 x 10^{24} formula units of SrO?

3. If a 15 g sample of $Na_2CO_3 \bullet n\ H_2O$ was heated until the water was boiled off, the mass of the anhydrate left would be 5.559 g Na_2CO_3. What is the n? What is the formula? What is the name of the hydrate?

Step #1: Find mass of substance without water.

Step #2: Find mass of water.

Step #3: Find moles of salt and water.

Step #4: Whole # ratio:

Step #5: Write the formula (and answer any other questions posed).

4. How many moles of yttrium are equal to 8.14 g of yttrium? (Assume molar mass is 88.9 g/mol)

5. What is the percent of water in $Na_2S_2O_3 \cdot 5\ H_2O$?

6. What is the molecular formula for a compound with a formula mass of 194 amu that contains 49.48% C, 5.15% H, 28.87% N and 16.49% O?
Step 1: Convert grams into moles

Step 2: Set up a mole ratio

Step 3: Turn the ratio into a chemical formula:
Step 4: Find the mass of the empirical formula

Step 5: Divide the molecule's mass by the empirical formula's mass. This will result in a small whole number.

Step 6: Multiply the empirical formula by this small whole number to obtain the molecule's formula.

Name: ________________________________ Date: _______________ Period: ______

KEY for Worksheet 13: Practice Test 1

Directions: Show all work. Use the conversion factor method where appropriate. Go to significant figures for conversions. Formula masses and molar masses, keep all decimals. Round percentages to 2 decimals. Short form is fine.

1. What is the molar mass of titanium (II) acetate ($Ti(C_2H_3O_2)_2$)?

1 Ti x 47.9 g = 47.9 g
4 C x 12.0 g = 48.0 g
6 H x 1.0 g = 6.0 g
4 O x 16.0 g = +64.0 g
165.9 g/mol $Ti(C_2H_3O_2)_2$

2. How many grams of strontium oxide are equal to 6.33 x 10^{24} formula units of SrO?

$$6.33 \times 10^{24} \text{ form u SrO} \times \frac{1\ mol\ SrO}{6.02 \times 10^{23} form\ u\ SrO} \times \frac{103.6\ g\ SrO}{1\ mol\ SrO} = 1090 \text{ g SrO}$$

Explanation: Start with the given, which is 6.33 x 10^{24} form u SrO. To get out of form u, 6.02 x 10^{23} form u SrO needs to be on the bottom part of the first conversion. 1 mol SrO will be on top of it. This will get you up to moles. To get from moles to grams, put 103.6 g SrO over 1 mol SrO. Solve by taking 6.33 x 10^{24} form u SrO and dividing by 6.02 x 10^{23} form u SrO, then multiply by 103.6 g SrO. The answer is 1089.348837 g SrO, but it becomes 1090 g SrO once rounded to 3 SF.

3. If a 15 g sample of $Na_2CO_3 \bullet n\ H_2O$ was heated until the water was boiled off, the mass of the anhydrate left would be 5.559 g Na_2CO_3. What is the n? What is the formula? What is the name of the hydrate?

Step #1: Find mass of substance without water. Given 5.559 g Na_2CO_3.

Step #2: Find mass of water. 15 g sample – 5.559 g Na_2CO_3 = 9.441 g H_2O

Step #3: Find moles of salt and water.

$$5.559 \text{ g } Na_2CO_3 \times \frac{1\ mol\ Na2CO3}{106\ g\ Na2CO3} = .05244 \text{ mol } Na_2CO_3$$

$$9.441 \text{ g } H_2O \times \frac{1\ mol\ H2O}{18.0\ g\ H2O} = .5245 \text{ mol } H_2O$$

Step #4: Whole # ratio:

Moles of Na_2CO_3 : Moles of H_2O

$$\frac{.05244}{.05244} : \frac{.5245}{.05244}$$

$$1 : 10$$

Step #5: Write the formula (and answer any other questions posed).
n = 10
Formula: $Na_2CO_3 \bullet 10\ H_2O$
Name: sodium carbonate decahydrate

4. How many moles of yttrium are equal to 8.14 g of yttrium? (Assume molar mass is 88.9 g/mol)

$8.14\ g\ Y \times \frac{1\ mol\ Y}{88.9\ g\ Y} = .0916\ mol\ Y$

5. What is the percent of water in $Na_2S_2O_3 \cdot 5\ H_2O$?
2 Na x 23.0 g = 46.0 g
2 S x 32.1 g = 64.2 g
3 O x 16.0 g = +48.0 g
158.2 g/mol $Na_2S_2O_3$
10 H x 1.0 g = 10.0 g
5 O x 16.0 g = +80.0 g
90.0 g/mol H_2O

Mass of Hydrate: 158.2 g/mol $Na_2S_2O_3$ + 90.0 g/mol H_2O = 248.2 $Na_2S_2O_3 \cdot 5\ H_2O$

% H2O in the hydrate: $\frac{90.0\ g/mol\ H2O}{248.2\ Na2S2O3 \cdot 5\ H2O} \times 100 = 36.26\%$

6. What is the molecular formula for a compound with a formula mass of 194 amu that contains 49.48% C, 5.15% H, 28.87% N and 16.49% O?
Step 1: Convert grams into moles

$49.48\ g\ C \times \frac{1\ mol\ C}{12.0\ g\ C} = 4.12\ mol\ C$

$5.15\ g\ H \times \frac{1\ mol\ H}{1.0\ g\ H} = 5.15\ mol\ H$

$28.87\ g\ N \times \frac{1\ mol\ N}{14.0\ g\ N} = 2.06\ mol\ N$

$16.49\ g\ O \times \frac{1\ mol\ O}{16.0\ g\ O} = 1.03\ mol\ O$

Step 2: Set up a mole ratio

Moles of C: Moles of H: Moles of N: Moles of O

$\frac{4.12}{1.03} : \frac{5.15}{1.03} : \frac{2.06}{1.03} : \frac{1.03}{1.03}$

4 : 5 : 2 : 1

Step 3: Turn the ratio into a chemical formula: $C_4H_5N_2O$

Step 4: Find the mass of the empirical formula

4 C x 12.0 amu = 48.0 amu
5 H x 1.0 amu = 5.0 amu
2 N x 14.0 amu = 28.0 amu
1 O x 16.0 amu = 16.0 amu
Mass of the empirical formula: 48.0 amu + 5.0 amu + 28.0 amu + 16.0 amu = 97 amu

Step 5: Divide the molecule's mass by the empirical formula's mass. This will result in a small whole number.

$$\frac{mass\ of\ the\ molecule}{mass\ of\ the\ empirical\ formula} = \frac{194\text{ amu}}{97\text{ amu}} = 2$$

Step 6: Multiply the empirical formula by this small whole number to obtain the molecule's formula.

2 x $C_4H_5N_2O$ = $C_8H_{10}N_4O_2$

Name: ______________________________ Date: ______________ Period: _____

Worksheet 14: Practice Test 2

Directions: Show all work. Use conversion factor method where appropriate. Go to significant figures for conversions. Formula masses and molar masses, keep all decimals. Round percents to 2 decimals.

1. What is the formula mass for $C_{21}H_{34}O_4$?

2. What mass of gold is contained in 6.88×10^{22} formula units of $AuBr_3$?

3. How many formula units of $Ti(C_2H_3O_2)_2$ are equal to 6.87 g of the compound?

4. If a 5 gram sample of $PbO_2 \cdot n\ H_2O$ was burned until all the water was gone, only 3.633 g of the anhydrate would be left. What is the value of n? The chemical formula for the hydrate and the name of the hydrate?

Step #1: Find mass of substance without water.

Step #2: Find mass of water.

Step #3: Find moles of salt and water.

Step #4: Whole # ratio:

Step #5: Write the formula (and answer any other questions posed).

5. What is the percent composition for potassium dichromate? (4 pts)

6. What is the empirical and molecular formula for the compound with a mass of 118 amu and a composition of 40.68% C, 5.08 % H, and 54.24% O?

Step 1: Convert grams into moles

Step 2: Set up a mole ratio

Step 3: Turn the ratio into a chemical formula (this is the empirical formula):

Step 4: Find the mass of the empirical formula

Step 5: Divide the molecule's mass by the empirical formula's mass. This will result in a small whole number.

Step 6: Multiply the empirical formula by this small whole number to obtain the molecule's formula.

Name: ______________________________ Date: _______________ Period: ______

KEY for Worksheet 14: Practice Test 2

Directions: Show all work. Use conversion factor method where appropriate. Go to significant figures for conversions. Formula masses and molar masses, keep all decimals. Round percentages to 2 decimals.

1. What is the formula mass for $C_{21}H_{34}O_4$?

21 C x 12.0 amu = 252.0 amu
34 H x 1.0 amu = 34.0 amu
4 O x 16.0 amu = +64.0 amu
350.0 amu

2. What mass of gold is contained in 6.88×10^{22} formula units of $AuBr_3$?

$$6.88 \times 10^{22} \text{ form u } AuBr_3 \times \frac{1\ mol\ AuBr3}{6.02 \times 10^{23} form\ u\ AuBr3} \times \frac{1\ mol\ Au}{1\ mol\ AuBr3} \times \frac{197.0\ g\ Au}{1\ mol\ Au} = 22.5 \text{ g Au}$$

3. How many formula units of $Ti(C_2H_3O_2)_2$ are equal to 6.87 g of the compound?

$$6.87 \text{ g } Ti(C_2H_3O_2)_2 \times \frac{1\ mol\ g\ Ti(C2H3O2)2}{165.9\ g\ Ti(C2H3O2)2} \times \frac{6.02 \times 10^{23} form\ u\ g\ Ti(C2H3O2)2}{1\ mol\ g\ Ti(C2H3O2)2} = 2.49 \times 10^{22} \text{ form u } Ti(C_2H_3O_2)_2$$

4. If a 5 gram sample of $PbO_2 \cdot n\ H_2O$ was burned until all the water was gone, only 3.633 g of the anhydrate would be left. What is the value of n? The chemical formula for the hydrate and the name of the hydrate?

Step #1: Find mass of substance without water. 3.633 g PbO_2

Step #2: Find mass of water. 5 g sample - 3.633 g PbO_2 = 1.367 g H_2O

Step #3: Find moles of salt and water.

$$3.633 \text{ g } PbO_2 \times \frac{1\ mol\ PbO2}{239.2\ g\ PbO2} = .0152 \text{ mol } PbO_2$$

$$1.367 \text{ g } H_2O \times \frac{1\ mol\ H2O}{18.0\ g\ H2O} = .0759 \text{ mol } H_2O$$

Step #4: Whole # ratio:

Moles of PbO_2: Moles of H_2O

$$\frac{.0152}{.0152} : \frac{.0759}{.0152}$$

1 : 5

Step #5: Write the formula (and answer any other questions posed).

n = 5

Formula: $PbO_2 \cdot 5\ H_2O$

Name: lead (IV) oxide pentahydrate

5. What is the percent composition for potassium dichromate? (4 pts)
Formula: $K_2Cr_2O_7$
2 K x 39.1 g = 78.2 g
2 Cr x 52.0 g = 104.0 g
7 O x 16.0 g =+112.0 g
294.2 g/mol $K_2Cr_2O_7$

$\% K = \frac{78.2 \text{ g K}}{294.2 \text{ g}} \times 100 = 26.58\%$

$\% Cr = \frac{104.0 \text{ g Cr}}{294.2 \text{ g}} \times 100 = 35.35\%$

$\% O = \frac{112.0 \text{ g O}}{294.2 \text{ g}} \times 100 = +38.07\%$

100%

6. What is the empirical and molecular formula for the compound with a mass of 118 amu and a composition of 40.68% C, 5.08 % H, and 54.24% O?

Step 1: Convert grams into moles

$40.68 \text{ g C} \times \frac{1 \text{ mol C}}{12.0 \text{ g C}} = 3.39 \text{ mol C}$

$5.08 \text{ g H} \times \frac{1 \text{ mol H}}{1.0 \text{ g H}} = 5.08 \text{ mol H}$

$54.24 \text{ g O} \times \frac{1 \text{ mol O}}{16.0 \text{ g O}} = 3.39 \text{ mol O}$

Step 2: Set up a mole ratio
Moles of C: Moles of H: Moles of O

$\frac{3.39}{3.39} : \frac{5.08}{3.39} : \frac{3.39}{3.39}$

1 : 1.5 : 1

2 x (1:1.5:1) = 2:3:2

Step 3: Turn the ratio into a chemical formula (this is the empirical formula): $C_2H_3O_2$

Step 4: Find the mass of the empirical formula
2 C x 12.0 amu = 24.0 amu
3 H x 1.0 amu = 3.0 amu
2 O x 16.0 amu = 32.0 amu
Mass of the empirical formula: 24.0 amu + 3.0 amu + 32.0 amu = 59 amu

Step 5: Divide the molecule's mass by the empirical formula's mass. This will result in a small whole number.

$\frac{\textit{mass of the molecule}}{\textit{mass of the empirical formula}} = \frac{118 \text{ amu}}{59 \text{ amu}} = 2$

Step 6: Multiply the empirical formula by this small whole number to obtain the molecule's formula.
2 x $C_2H_3O_2$ = $C_4H_6O_4$

Thank You for Reading:

Chemistry takes a lot of practice. This workbook focused on Mole Concepts. Check out *5 Steps to Surviving Chemistry* if you're looking for a more comprehensive review of other chem topics. You may also wish to grab *5 Steps Chemistry Workbook Series Book 1: Sassy Stoichiometry Concepts*.

I hope this book helps you along your journey. If you have questions, please feel free to email me at devyaschildren@gmail.com.

If you're ready for a break from chem, visit my website (http://www.juliecgilbert.com/) to read up on some of my fiction series. Whether you're in the mood for a heavy hitter like Beyond Broken Pencils (about a school shooting), an epic scifi series like Anotech Chronicles, or a light mystery/thriller like the Megan Luchek series, there's probably a story that will appeal to you.

If you're interested in virtual tutoring via zoom, send inquiries to the above email address.

Good luck.

Sincerely,

Julie C. Gilbert

Notes Page 1:

Notes Page 2:

Notes Page 3:

Made in United States
North Haven, CT
06 November 2026

21793611R00043